Deep, Dark Secrets Of Black Boxing

Delivering a Usable Spec For An IP

Edward Seymour

Deep, Dark Secrets Of Black Boxing

Delivering a Usable Spec For An IP

Edward Seymour

This text is dedicated to revealing a new paradigm for spec wrting
in electronics that is time tested and proven

Table of Contents

About the author

I began my study of computer architecture with intel 8085 and motorola 68000 long ago. My studies continued within IBM with the IBM PC, 370 VMS and VM operating systems and even involved a study of pdp 11/44 from digital equipment corporation. From there began a journey with PowerPC architecture. Throughout this time it became increasingly important to incorporate design which came from others.

In some cases, the source of design reuse was another division of IBM. In other cases it was a vendor providing a "component". From a Physical Design prospective, this widget always had a box with pins and some obstructions on certain metal layers (hard IP) or it had some loosely defined logic function, input to synthesis (soft IP).

The overriding usage model was, just like the TI Databook. It had a picture of some widget, boundary spec for range of voltage/temperature/speed, it was guaranteed to operate. Perhaps it may operate outside these constraints but this was buyer beware.

As applied to timing analysis, the box had a prescribed transfer function expressed from input to output to confirm when you drop it into a design, it works, end of story.

Functionally, it had a behavioral model to represent the intended operation. Essentially, it has inputs and outputs. In my career I have built specs, built circuits in any possible way, tested the result and had success, along with sporadic failure.

Background

The aforementioned tenets have remained unchanged since 1972 when I first built circuits.

Crux Of My Idea

Conveyance of ANY IP EVER should be expressed in TI Databook form where you have a single Top Level Box. The Box has pins (all inclusive of Power, Ground, Signals, etc).

For all pins on the device, express all modes of operation in a single table with columns or rows for each defined fucntion.

There must be a timing model which covers all modes of operation. It can be extracted to summarize or simply, a file, usable by any commercial timing tool.

For hard IP, there needs to be a physical abstract (box with pins) germane to the technology at hand. For soft IP there needs to be an RTL representation in a file or files provided.

Paper budget for all the above, excluding the aforementioned files is 3 pages, font size >=10.

Concrete Examples of Where This Worked

Undergradute Engineering and Computer Science Courses

In IBM when I had the task to teach chip design and test on campus, I first needed to write textbooks on all related topics that relayed IBM Secrets in a way they became public knowledge. This was a daunting task for a junior engineer but I created these "specs" and taught classes for 7 years.

Papers created in this program across 4 universities complied with the rules of specification in conveying designs. These were 4.2mm x 4.2mm chips 1983-1989. Details can be found in my paper entitled IBM VLSI Academic Program.

Automotive Electronics - Chrysler

Consulted at Chrysler Jeep Electronics while working for IBM Marketing headquarters to help prescribe methods to vastly improve quality and time to market (1990)

Automotive Electronics – Ford

Had a consulting practice and Ford in Dearborn Michigan to offer Ford the same advantage offered Chrysler. I did this without disclosure of any details from Chrysler. Once again, IBM secret methods applied outside IBM, for the benefit of consumers.

PowerPC Use By Apple And Groupe Bull

Same rule as aforementioned applied to the IP called PowerPC from IBM, used by Motorola and Apple. I have patents on test engines used, ran the system bringup lab and debugged product introduction on Europe's first 64 bit Unix Machine in 1996.

Power PC in AMD

At this juncture, I conveyed the function of processor cores to the AMD effort. Physical design and timing were deliverables. Same methods of conveying specification was routinely used here. This included consultation on AMD running our process and the sharing techniques.

Power4 – Hotchips Presentation

Story of how Power4 worked for physical design and test. It took very few slides I still have.

Power 5 – Core that made the Apple Cutting Edge Box

Delivered a wide bandwidth core to a 40 pin interface. It broke records in performance but took lots of power.

Power 6 – Beat Kasperov in Chess

Did array and logic testing for this. Reviewed cermamic package for test and performance. Specified test board. Implemented yield learning process.

Power 7 – 32 nm – Beat Jeopardy Champions

Did process bringup, processor debug, found and fixed design,
process interactions, again.

Power 8 – 22nm

Same as power7 effort with techniques for yield learning

14nm and beyond

Improved yield learning process, sale to Global Foundries

Customer List of > 50 companies

Black Box Axioms

Understanding of Cradle To Grave

Fundamental Grasp of Customer Culture

Detailed Plans That Deliver On Time, Under Budget With Excellence

No Patience For A Person Who Refuses To Do Math

Insistence on Scientific Method

Critical Of Any Analysis Which Fails To Define Measurement Units and Precision

If You Have Something Elegant, It Should Take No More Than Three Pages to Convey

A Taste Does Not Require Disclosure of Recipe

If There Are Too Many Words, Message Is Lost

Just Like A Black Box Recorder, Has To Be Hooked Up Exactly

True Sign Of Science, It Can Be Easily Taught

Follows Engineering Principals

Employs Scientific Method